ELECTRICITY & MAGNETISM

Explorations in Electricity & Magnetism

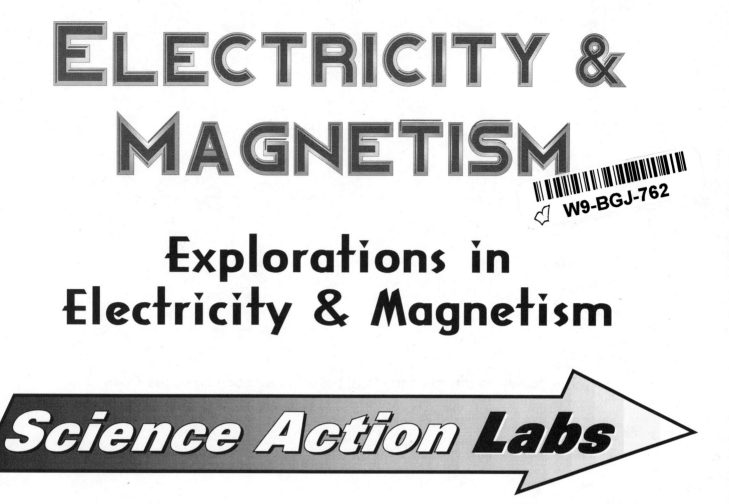

Science Action Labs

Written by Edward Shevick

Illustrated by Leo Abbett

Teaching & Learning Company

1204 Buchanan St., P.O. Box 10
Carthage, IL 62321-0010

This book belongs to

_____

The activity portrayed on the front cover is described on pages 44-45.

Cover design by Kelly Harl

Copyright © 2000, Teaching & Learning Company

ISBN No. 1-57310-207-5

Printing No. 98765432

Teaching & Learning Company
1204 Buchanan St., P.O. Box 10
Carthage, IL 62321-0010

Table of Contents
Science Action Labs

Dear Teacher or Parent,

The spirit of Sir Isaac Newton will be with you and your students in this book. Newton loved science, math and experimenting. He explained the laws of gravity. He demonstrated the nature of light. He discovered how planets stay in orbit around our sun.

Electricity & Magnetism can help your students in many ways. Choose some activities to electrify your class demonstrations. Some **electricity and magnetism units** can be converted to hands-on lab activities for the entire class. Some can be developed into student projects or reports. Every class has a few students with a special zest for science. Encourage them to pursue some activities on their own.

Enjoy these science experiments as much as Newton would have. They are designed to make your students **think**. Thinking and solving problems are what science is all about. Each **electricity and magnetism unit** encourages thought. Students are often asked to come up with their best and most reasonable guess as to what will happen. Scientists call this type of guess a **hypothesis**. They are told how to assemble the materials necessary to actually try out each activity. Scientists call this **experimenting**.

Don't expect your experiments to always prove the hypothesis right. These science activities have been picked to challenge students' thinking abilities.

All the activities in **Electricity & Magnetism** are based upon science principles. That is why Sir Isaac Newton has been used as a guide through the pages of this book. Newton will help your students think about, build and experiment with these activities. Newton will be with them in every activity to advise, encourage and praise their efforts.

The answers you will need are on page 64. You will also find some science facts that will help your students understand what happened.

Here are some suggestions to help you succeed:

1. Observe carefully.
2. Follow directions.
3. Measure carefully.
4. Hypothesize intelligently.
5. Experiment safely.
6. Keep experimenting until you succeed.

Sincerely,

Ed

Ed Shevick

TLC10207 Copyright © Teaching & Learning Company, Carthage, IL 62321-0010

Name _____

Magnets in Our World

Newton Explains Magnets

Lodestone

Cork

The earliest magnets were natural magnets called lodestones. They were rocks that had magnetic qualities. Ancient Chinese and Greeks studied them. They believed that lodestones had medical uses. They even discovered that a lodestone mounted on a cork would float in water and act as a compass.

Try to obtain a lodestone. Experiment to see if it can pick up a pin or small paper clip.

Today there are many kinds of magnets. Some are made of iron or steel. Some are made from ceramics mixed with iron filings. The best magnets are made of an alloy called **alnico**. This name tells you that these powerful magnets contain **al**uminum, **ni**ckel and **co**balt.

KINDS OF MAGNETS

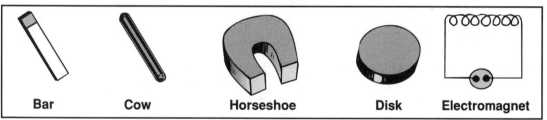

| Bar | Cow | Horseshoe | Disk | Electromagnet |

You may be wondering about the "cow" magnets. They are powerful magnets about the size of your little finger. Farmers have cows swallow them to get rid of the wire they may have taken in with baled hay.

Magnets Are Useful

You wouldn't have electricity in your home without magnets. Your television has many magnetic parts. Without magnets your telephone would not work.

On the next page is a list of devices that use magnets. The letters are scrambled. Can you unscramble them?

Name _____

1. dirao _____   2. morots _____

3. passcom _____   4. darar _____

5. lebsl _____   6. srekaeps _____

7. yots _____   8. temers _____

9. pate _____   10. epcuotmr _____

There are hundreds of other uses for magnets. List some magnets that you might have at home, school or work.

1. _____

2. _____

3. _____

4. _____

5. _____

Magnet imagination time. Can you come up with two novel uses for magnets? Your imaginary magnets can be any size, shape or color you wish. A sketch might help.

1. _____

_____

_____

2. _____

_____

_____

Magnetic Attraction

Magnets attract some objects. Magnets have no effect on other objects. Let's experiment to find what materials respond or do not respond to a magnet.

Magnets in Our World

Name _____

1. Obtain a magnet and a box full of various materials from school or home. Your home workshop or garage may have a wide variety of materials. Don't forget coins.

2. Try picking up various objects with your magnet, and fill out the Magnet Data Table. Check the correct column for each material.

MAGNET DATA TABLE		
Kind of Material	**Attracted by a Magnet**	**Not Attracted by a Magnet**
Example: Dollar bill		✓
1.		
2.		
3.		
4.		
5.		
6.		
7.		
8.		
9.		
10.		
11.		
12.		
13.		
14.		
15.		

3. Look over your data table. What **type** of materials are attracted by magnets?

   _____

4. You may have answered *metals* in the above question. What kind of metals were **not** attracted? _____

5. You made a goof and mixed thumbtacks in sand. What would be the easy way to separate the tacks and sand? _____

Newton Note: You have discovered that magnets mainly attract materials containing iron. Scientists call these ferrous.

Name _____

Magnetic Passage

Now let's ask your magnet another question. What materials will let the magnet forces go through them?

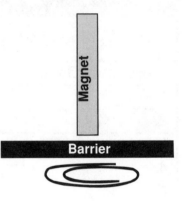

1. Obtain various materials that may act as a barrier to magnetism. Use paper, cardboard, wood, cloth, aluminum foil, glass, plastic, iron sheets, water or ?

2. Try to attract a paper clip through various barriers. List the materials that let the magnetism through.

   _____

   _____

   List the materials that did not let the magnetism

   through. _____

3. Suppose magnetism was discovered to be harmful. What material would you

   make clothes of to protect yourself from magnetism? _____

Making and Testing Magnets

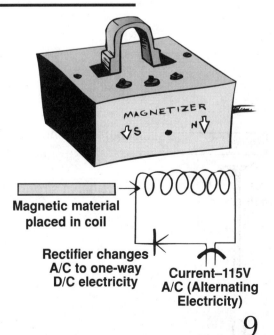

NEWTON'S
ACTION LAB
Electricity &
Magnetism
2

Newton Explains Magnet Theory

Magnets are made of atoms. The atoms of magnetic materials act like miniature magnets. Each has a north and south pole. Sometimes a number of atoms combine to form magnetic areas called **domains**.

You can't see atoms or even domains. Here is how scientists think they appear in magnets. Strong magnets have most of the domains lined up.

Not Magnetized

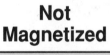

Domains facing
in all directions.

Weak Magnet

Only some
domains lined up.

Strong Magnet

Practically all
domains lined up.

Making Magnets

Magnets can be made by placing magnetic materials in a strong field of **direct** current electricity. The current moving in **only one** direction lines up the domains.

Here is how the magnetizer works: regular 115 volts AC electricity is fed into a coil of wire. The rectifier changes the AC into the DC or one-way electricity needed to magnetize. The magnetic material is placed in the coil where its domains are lined up.

Magnetic material placed in coil

Rectifier changes A/C to one-way D/C electricity

Current–115V A/C (Alternating Electricity)

Name _____

There is a simpler way of making a magnet if you already have a strong magnet. This method involves stroking magnetic material across the good magnet.

1. Obtain a strong bar magnet, a few thumbtacks or paper clips and a 3" (8 cm) long, thick nail.

> **Caution!** Magnets can harm watches. Remove all watches during magnet experiments. Also keep magnets away from televisions, computers and telephones.

2. Stroke the nail along the magnet as shown about 20 to 30 times. Always stroke the nail in the **same** direction. When you get to the end of the magnet, lift the nail **away** from the magnet. Then move the nail back, near the front of the magnet and repeat the stroke. See the diagram on the left.

3. Now use your nail to pick up some paper clips or thumbtacks. Congratulations! You have made a magnet. Do the domains inside your magnetized nail look like **a** or **b** below?

A

B

> **Newton Hint:** Nails are made of "soft" iron. They lose their magnetism easily. Try knocking your magnetized nail against a safe object. You will be mixing up the magnetic domains. Now check the nail to find if it has lost some of its magnetism.

The Poles of a Magnet

Magnets are not equally strong all over. Let's experiment to find out where a magnet is the strongest.

1. Obtain a strong bar magnet, a 2" or 3" (5 or 8 cm) long nail and 10 paper clips.

2. Place the nail head as shown. The nail head should be 1" (2.5 cm) away from either end. What happened to the nail? _____

_____

one inch

Name _____

3. Move your nail head from one end of the magnet to the other. Describe what you **feel** at the ends and center. _____

4. Where is a magnet strongest? _____

5. Where is a magnet weakest? _____

6. Line up 10 paper clips as shown. They should be lined up so that they have the same length as the magnet.

7. Bring your magnet down toward the lined-up paper clips. What happened to the end paper clips? _____

8. What happened to the middle paper clips? _____

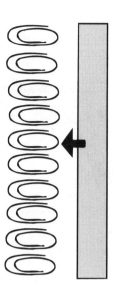

Newton Note: A magnet's poles are stronger because that is where the "lines of magnetic force" are greatest. The next lab will explain lines of force.

How Magnets Repel and Attract

You've learned that magnets have poles at each end. These poles are the strongest parts of a magnet.

There are two kinds of poles on a magnet. One is called north and the other south. The north end of a suspended magnet will point toward the north direction of the Earth. Let's find out which end of our bar magnet is the north pole.

1. Obtain a strong bar magnet, string and a **large** paper clip.

2. Use pliers to construct the cradle as shown. It should be a little shorter than your magnet. It is also possible to make a cradle of string tied to the magnet.

3. Balance the bar magnet in the cradle, tie a string to it and tape the string end to a table edge.

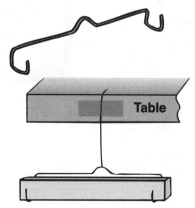

4. Leave the suspended magnet alone as it faces **almost** north and south.

5. Use masking tape to mark the north pointing end as a north pole. Now you are ready to find out how magnet poles attract and repel.

Name _____

6. Obtain a second bar magnet.

7. Bring one end of the second bar magnet slowly toward the north pole of your **suspended** magnet. If the second bar magnet attracted the north pole of the suspended magnet, use tape to mark it as a south pole. If it repelled the suspended magnet, mark it as a north pole.

Now you have identified both poles of both magnets.

8. Bring the **south** pole of the magnet **slowly** toward the **south** pole of the suspended magnet.

Describe what happened. _____

_____

9. Bring the **north** pole of the magnet to the **south** pole of the suspended magnet.

Describe what happened. _____

_____

10. Let's summarize what you have learned by filling in the blanks below.

    Like poles (such as north and north or south and south) _____ each other.

    Unlike poles (such as a north and south pole) _____ each other.

11. Place two bar magnets flat on a table. Slide various poles together to demonstrate attractions and repulsion.

12. Hold a magnet in each hand to "feel" attraction and repulsion.

Maglev Trains. *Maglev* is short for **magnetic lev**itation. Maglev trains use magnetic repulsion to raise themselves a very small distance above the tracks. This allows less friction and faster trains. Both England and Japan have Maglev trains.

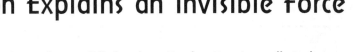

Magnetic Lines of Force

Newton Explains an Invisible Force

You throw a ball into the air. It always falls back to Earth. Gravity pulls it down. Gravity is a strong **invisible force.** You know it is there, but you never see it.

Magnetism is also an invisible force. Every magnet has an invisible force around it called **lines of force.** The stronger the magnet, the stronger are the invisible lines of force.

Our Earth acts like it has a giant magnet inside. The Earth's hidden magnet does not line up with the north and south poles that form its spinning axis. The magnetic north pole is in northern Canada. The magnetic south pole is near the coast of Antarctica.

We will be testing the Earth's magnetism in the compass lab that follows this lab. Let's use iron filings to help us "see" the lines of force around magnets.

North Pole

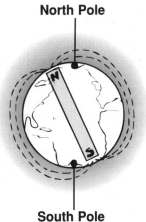

South Pole

What Is Around a Bar Magnet?

1. Obtain a bar magnet and a saltshaker full of iron filings. You can use small thumbtacks instead of iron filings, but the filings work much better.

2. Obtain a manila file folder or any 8$^1/_2$" x 11" (22 x 28 cm) piece of cardboard or plastic.

3. Place the folder on the table. Place the bar magnet in the center of the folder and sprinkle iron filings all around the magnet.

4. Tap the folder gently with your finger to get the best view of the lines of force.

Here is what you should see. Are your lines of force greater at the ends or middle of the bar magnet? _____

Caution! Clean up and save your iron filings. It won't be easy removing them from the magnet.

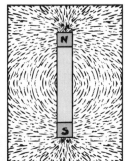

Name _____

More Lines of Force

Now let's try other lines of force pictures. Again sprinkle iron filings on the magnets arranged as below. The magnets should be 1$^1/_2$" (4 cm) apart. Don't forget to tap gently and clean up carefully. Draw the patterns you find.

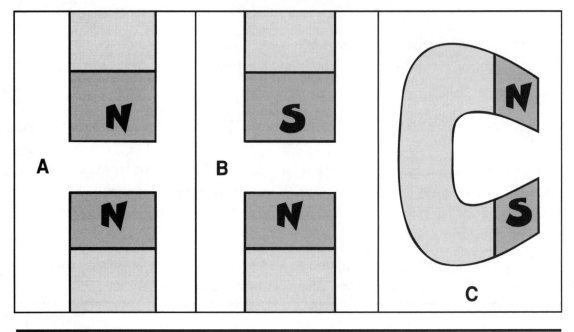

Lines of Force Freedom

Try other magnet combinations. Try placing an iron washer in between the poles. Magnetize a nail by stroking it one way along a strong magnet and view its lines of force. Sketch your arrangements and the iron filing patterns below.

14

Compass Lab

The Earth as a Magnet

Our Earth is a giant magnet. This is due to the slight motion of the iron-nickel molten core at the Earth's center. The result is magnetic lines of force around the Earth. Like any other magnet, the Earth has a magnetic north and south pole.

Confusion time. The magnetic north and south pole are not located at what is called the **geographic north and south pole**. The geographic north and south pole form an axis about which the Earth rotates. The magnetic north pole moves over time. The magnetic north pole is located near Hudson Bay in Canada. The magnetic south pole is near the Antarctic coast.

North Magnetic Pole   **North Geographic Pole**

South Magnetic Pole   **South Geographic Pole**

Using a Compass

When you use a compass to find your way, you are using the Earth's magnetic lines of force. The magnetic lines of force of your compass needle react with the Earth's magnetic field to point to the north and south magnetic poles. In today's lab you are going to learn how to read and use a compass.

Obtain a compass and study its face. Most simple compasses look like the sketch to the right. Here is how to use your compass.

1. Hold the compass waist high in your hand.

2. Hold it as level as you can.

3. Rotate your **entire body** to line up the north on the needle with the *north* printed on your compass.

4. Both your needle and your hand will not be steady. Make the best guess you can on the angle of the objects you find. See the tree example.

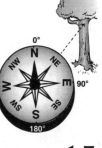

Name _____

5. From your desk describe *exactly* what part of the room is in each of the following directions. Remember to keep your compass away from the metal of the table, hold it level and rotate the compass until the needle is aligned North-South.

   1. North _____   **2.** East _____

   3. Southeast _____   **4.** South _____

   5. West _____   **6.** Northwest _____

6. Line up your compass as before. Give the approximate *degree* bearing from your position to the areas in the room that your teacher has marked with signs numbered 1 to 6.

   1. _____°   2. _____°   3. _____°

   4. _____°   5. _____°   6. _____°

Compass Treasure Hunt

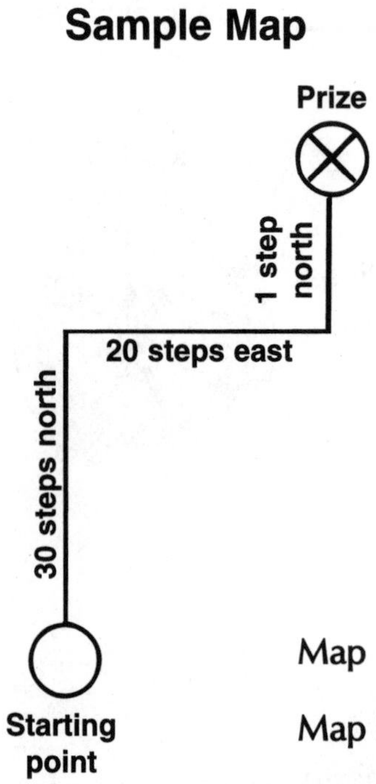

Sample Map

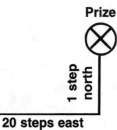

Prize

1 step north

20 steps east

30 steps north

Starting point

Your compass can help you find treasure. Your teacher has hidden prizes around your school. They may be gold, candy or free points on a future exam. You will be given the compass direction and number of steps to each prize. You may be sent in zigzag directions for each prize.

Copy all four treasure maps from the board. Your team will be assigned one of them and a starting point. Following directions exactly, find the point marked *X* on the map. Remember, other classes are in session and you must be on your best behavior. There will be no running, shouting or random hunting. Under no circumstances are you to do anything unsafe or go on the street. Follow the orders of the students who have set up the treasure hunt. In the spaces provided describe where the point marked *X* is on your treasure map.

Map 1: _____

Map 2: _____

Map 3: _____

Map 4: _____

16

Name _____

Making Your Own Compass

Here are two simple primitive compasses that you can build. Both use a sewing needle that has been magnetized by being stroked along **one** pole of a strong magnet.

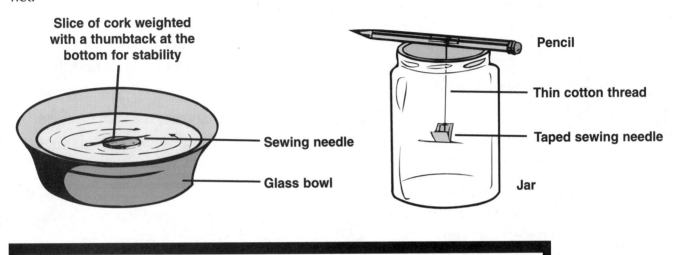

Slice of cork weighted with a thumbtack at the bottom for stability

Sewing needle

Glass bowl

Pencil

Thin cotton thread

Taped sewing needle

Jar

Gyrocompass. Modern sailing ships do not rely on magnetic compasses. Magnetic compasses are affected by the ships' metal and do not point true north. A gyrocompass uses one or more spinning gyroscopes (wheels) to keep track of true geographic north.

Electromagnet Lab

Newton Explains Electromagnets

Iron core
Switch

Battery   Coil of wire

Magnets are used in thousands of electrical devices. The most useful kind of magnet is **not** the bar or horseshoe magnet. Bar and horseshoe magnets are **permanent** magnets. They cannot be turned off.

Electromagnets can be turned on and off. They are basically a coil of wire with an iron core. When electricity goes through the wire, it becomes a strong magnet. When the electricity is turned off, it loses its magnetism.

Study the top diagram on the left. It shows a typical electromagnet. Closing the switch allows electricity to make it a magnet. When you open the switch, there is no electricity and no magnetism.

A bell is a good example of an electromagnet at work. The push button at your door is the switch. It sends electricity to an electromagnet in the bell. This causes a clapper to strike the bell, and you know you have a visitor.

Doorbell

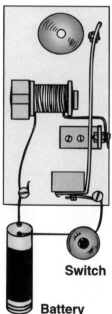

Switch

Battery

Making Your Own Electromagnet

Here is how to make your own electromagnet.

1. Obtain a six-volt battery, about 6' (1.8 m) of plastic covered wire, a large 5" or 6" (13 or 15 cm) nail and a few small paper clips.

Caution! Electricity going through wires generates heat. Handle the wires carefully. **Do not** leave the wires attached to the battery too long. Disconnect the battery as soon as you get results.

2. Start winding your electromagnet at the head of the nail. Leave 6" (15 cm) of wire extending out for the battery connection. (See next page.)

3. Wind the coil very tightly with each coil of wire touching the next.

4. Leave another 6" (15 cm) at the other end of the coil to connect to the battery.

Name _____

5. Now it is time to test your electromagnet. **Remember not to leave the battery connected for too long.**

6. Connect the battery and try to pick up a paper clip.

Describe what happened. _____

_____

6 inches of wire

6 volt BATTERY

7. Disconnect your battery and try to pick up a paper clip.

Describe what happened. _____

_____

8. Try picking up bigger and bigger metal objects such as nails, ball bearings or ?

9. What was the heaviest object that your electromagnet picked up? _____

Electricity and Magnetism

Electricity and magnetism are related. Any wire carrying electricity has magnetic lines of force around it. Magnets moving in a coil of wire can generate electricity.

1. Reassemble the electromagnet that you used before.

2. Obtain a simple compass.

3. Place your compass below your electromagnet.

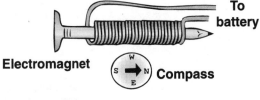

To battery

Electromagnet   S ➡ N **Compass**

4. What happens to the compass needle as you connect the electromagnet to

the battery? _____

5. What must be surrounding an electromagnet to cause the compass needle to

jump? _____

Newton Explains How to Make Electricity

Most of the electricity we use in our homes is generated by using wire coils and magnets. All you need is a source of energy to move either the magnet or the coil rapidly.

Name _____

Generating electricity is not as simple as the diagram shown below. Some electric generators are as big as your home.

If you can obtain an electric meter, try to generate some electricity. Wrap 30 coils of wire around the cardboard tube of a roll of toilet paper. Hook it to the meter and plunge a bar magnet into the coil.

Newton Challenge: Electromagnets are very useful. They help to start your automobile. They are part of burglar alarm. Scrap yards use them to load and unload iron metals.

Can you think of a **novel** use for an electromagnet? You don't have to build a model. Just dream it up and describe and sketch your novel idea below.

_____

_____

_____

_____

_____

Generating Electricity

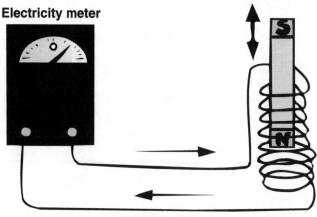

Electricity meter

Magnet moving
in and out

Electron flow

Millimag Lab: Investigating Magnetic Strength

Newton Introduces the Gauss

Scientists measure everything. They measure electric strength in volts. They measure electric current in amperes. They even measure the strength of a magnetic field in a unit called **gauss**.

The magnetic field around the Earth is very weak. It is only $1/4$ gauss at the equator and $3/4$ of a gauss at the North Pole. A toy magnet can measure 1000 gauss while a powerful electromagnet can be over 30,000 gauss.

Our lab does not have a gauss meter, so we are going to invent our own unit. It will be called a **millimag**. This will be the distance in millimeters at which a magnet can begin to attract a right angle bracket. Our unit will not relate directly to the gauss, but it will allow us to compare the strength of various kinds of magnets.

> **Newton Note:** The unit gauss is named after Karl Friedrich Gauss (1777-1855). Karl Gauss worked on the math of electromagnetism. He was a child genius. By 19 years of age, he was world famous for his math theories.

Millimag Measurements

1. Obtain two bar magnets, one horseshoe magnet and any other magnets you can find from toys, speakers and elsewhere. Refrigerator magnets work well. Share magnets with other lab teams.

2. Obtain a metric ruler and a 1" (2.5 cm) right angle metal bracket.

3. Study the metric ruler. You will be using the smallest units called **millimeters**. There are 10 millimeters to every centimeter.

Name _____

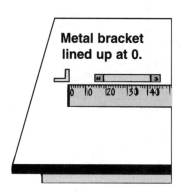

Metal bracket lined up at 0.

4. Tape your ruler to a table as shown.

5. Place your bracket at the "0" mark and at the side of your metric ruler.

6. **Slowly** bring your bar magnet along side of the ruler until the bracket is attracted. Note the millimeters at the point of attraction.

7. Repeat a few times till you are sure of the millimeter distance. Then record the millimeters in the Millimag Data Table.

8. Repeat using the other bar magnet, the horseshoe magnet and any other available magnets.

9. Record each millimeter reading in the data table.

MILLIMAG DATA TABLE

Magnet Used (Describe It)	Millimeter Attraction
Example: Toy Magnet	12
1. First bar magnet	
2. Second bar magnet	
3. Horseshoe magnet	
4.	
5.	
6.	
7.	
8.	

Graphing Your Millimag Data

A good bar graph can help organize and make sense of your data. Place your millimag data on the bar graph on the following page. Be neat and add color to your bars.

Name _____

MILLIMAG BAR GRAPH

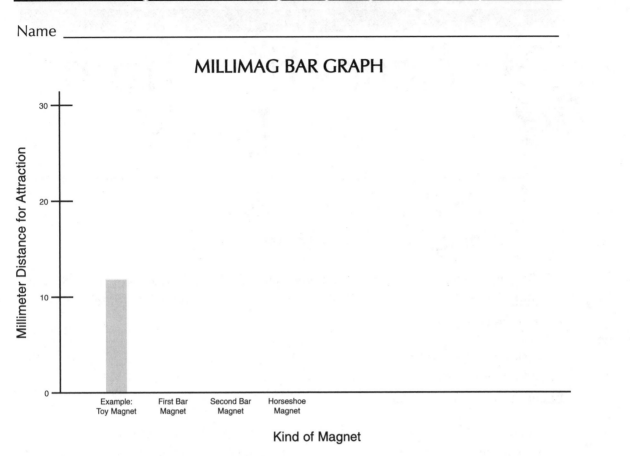

Millimeter Distance for Attraction

Kind of Magnet

Example: Toy Magnet   First Bar Magnet   Second Bar Magnet   Horseshoe Magnet

What Can Magnets Pass Through?

Gravity forces can pass through almost anything. Magnetic lines of force go through many materials. Some materials stop magnetic lines of force. Let's find which material can or cannot stop magnetism.

1. Use the same setup as you did for millimag measurements.

2. Collect all kinds of material to be tested. Try plastic, cloth, wood and glass. Try metals such as iron, steel, aluminum and copper. Even try using the flesh of your finger as a barrier.

3. Place the barrier right in front of the magnet as you move it forward. Construct your own data table to record the millimeter attraction distances. If you are ambitious, construct a bar graph with your data.

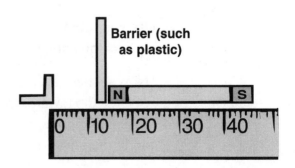

Barrier (such as plastic)

Magnetism from the Earth

Newton Explains How Magnets Are Made

Lodestone

We live on a magnet. Our Earth is a very large but very weak magnet. Like any magnet, the Earth is surrounded by lines of force. It is these lines of force that move the needle of a compass.

Parts of the Earth are magnetic. Some iron-bearing ore called **lodestones** are magnetized by the Earth's magnetism. Lodestones mounted on corks floating in water became the world's first compasses.

Magnetizing a nail

You can make a magnet by stroking iron or steel (such as a needle, nail or open paper clip) near a strong magnet. You must stroke in the **same** direction along only **one** pole of the magnet.

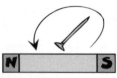
Direct current electricity

Real magnets are made with steel or magnetic alloys. They are placed inside a coil of wire through which direct current is sent.

There is still another way to make a magnet. Read on.

Magnetism with a Hammer

1. Obtain a hammer, a compass and an iron or steel rod about 2' (0.61 m) long. Iron curtain rods work. Check your local hardware store for rods.

2. Use your compass outdoors to locate magnetic north. Make north with a line or stick pointing north.

Measuring Earth's tilt

Now we have a small problem. We have to find out the angle at which the lines of force dip towards the Earth. Scientists use a dipping needle to find this angle. A simple dipping needle can be made by magnetizing a steel knitting needle and mounting it on a cork as shown. Use two small needles in a cork for support. There is an easier way to find the angle of dip.

The angle of dip is called **magnetic inclination**. On the following page are the approximate angles as published in books.

Name _____

North Magnetic Pole . . . . .90°	Illinois .70°
Equator . . . . .180°	California . . . . .60°
Puerto Rico . . . . .40°	Florida .57°
Hawaii . . . . .39°	The average magnetic dip in the
New York . . . . .74°	United States is about . . . . .65°

Now it's time to magnetize your metal rod.

1. Face your metal rod toward north.

2. Tilt it at about 65°. Place some tape at the top end. You'll need to identify the top later.

N ←

65°

3. Strike the top end of the rod sharply and **carefully** about 10 times with the hammer.

Testing Your Metal Rod for Magnetism

According to magnetic theory, the top end of your rod has become a south pole. The bottom end has become a north pole. Here is how to test your rod.

1. Bring the top end of your rod (the south pole) toward the north needle of your compass. It should be attracted.

What really happened? _____

2. Bring the top end of your rod to the south end of your compass. It should repel.

What really happened? _____

3. Repeat facing the bottom end of your rod (the north pole) to both needles on your compass.

Describe what happened. _____

_____

Newton Note: Magnetizing a rod with a hammer and the Earth is not easy. Sometimes the rod's magnetism is too weak to detect. Try other rods, more hammer blows or any other variation.

Name _____

Magnetic Challenge

The Bar Magnet Puzzler

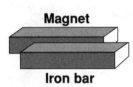

Magnet

Iron bar

You have a bar magnet and an iron bar. Both are the same size, shape, weight and color. There are no visible differences between the bar magnet and the non-magnetized iron bar. How can you tell which is the magnet and which is the iron bar? **You cannot use anything but the magnet and iron bar to determine which is which.** You **cannot** use paper clips, strings to suspend the bars, tools or scientific instruments.

Think! This is a solvable **mental** problem if you know something about magnets. It is highly unlikely that you will really find a magnet and a similar non-magnetic bar. Write your best ideas below.

_____

_____

_____

Newton's Magnetic Car

Here is a magnetic challenge for you. If you could make this work, you will become rich and famous.

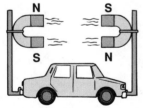

Newton mounted a powerful horseshoe magnet above the front bumper of his car. He mounted another powerful magnet above the rear bumper. He faced the magnets so they attracted each other. As the front magnet pulled on the rear magnet, the car was supposed to zoom forward.

Newton failed to create a car that would run on magnetism instead of gasoline. Try

to explain why Newton's magnetic car failed. _____

_____

_____

Name _____

Paper Clip Pick Up Challenge

Challenge your classmates to a paper clip pick up contest. The winner is the one that can attract and hold the longest string of paper clips.

Challenge 1

Everybody uses the **same** bar or horseshoe magnet. Everybody uses the approximately 1" (2.5 cm) long paper clips.

What was the largest number of suspended clips? _____

Challenge 2

Everybody uses the **same** small paper clip, but students are allowed to find the strongest magnet that they can.

What was the largest number of suspended clips? _____

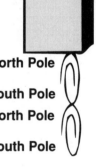

North Pole

South Pole

North Pole

South Pole

> **Newton Note:** Each suspended paper clip actually becomes a magnet with a north and south pole. That is why they remain suspended until the last magnetized paper clip is too weak.

Vending Machine Magnets

Vending machines for candy or soft drinks usually have a magnetic device. This is to keep dishonest people from using steel washers instead of non-magnetic coins. The magnet pulls the slugs to one side so that they don't operate the vending machine.

Here is your magnetic challenge. Can you make a slug rejector?

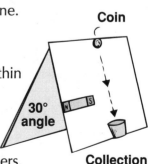

Coin

Collection cup

1. Build a simple 30° ramp for the slugs and coins to slide down. You could use thin wood or cardboard.

2. Collect various coins and iron or steel washers.

3. Attach a magnet halfway down the ramp in a position that deflects the iron washers.

4. Place a paper cup at the base to catch the real non-magnetic coins.

   Enjoy your slug rejector.

Name _____

Magnetic Fun

Magnetic Levitation

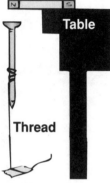

Table

Thread

Tape to floor

To *levitate* means "to raise up." Magicians often levitate an assistant. The assistant rises up into the air with no visible means of support.

Magnetism can be used to levitate a nail or paper clip.

1. Obtain a magnet, nail or paper clip, light thread and tape.

2. Tape the magnet to the table so that it extends beyond the edge.

3. Tie the thread to a nail and hook up your levitation experiment as shown.

Here is another way to demonstrate magnetic levitation.

4. Obtain a board **roughly** 12" (30 cm) long by 6" (15 cm) wide and $^1/_2$" (1.25 cm) thick.

5. Obtain two strong bar magnets, a hammer and six long aluminum or brass nails. **You cannot use iron nails.**

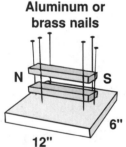

Aluminum or brass nails

N   **S**

12"   **6"**

6. Build the device shown to the left. The nails should be spaced to loosely contain the magnets.

7. Place the bar magnets with north facing north as shown. Push the top magnet down with your finger.

Describe what happened. _____
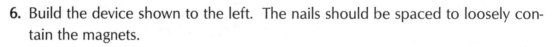

8. Brainstorming time. Can you dream up some devices that could use this levitation concept? For example, you could use it to replace springs in a bed. List

   your ideas. _____

A Money Magnetic Maze

Magnets can help you deposit money in our bank of magnetism. All you need are pennies, dimes, small paper clips and a strong magnet.

Name _____

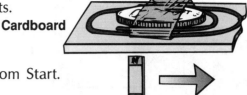

Paper clip taped to coin

Cardboard

1. Attach your pennies and dimes to a small paper clip using tape. You need paper clips because coins are not attracted to magnets.

2. Place the maze below on some stiff cardboard.

3. Use the magnet as shown to move your money away from Start. Finish at the Bank of Magnetism.

4. Try timing how long each student takes to deposit three pennies and three dimes. Have all the coins attached to paper clips before you start.

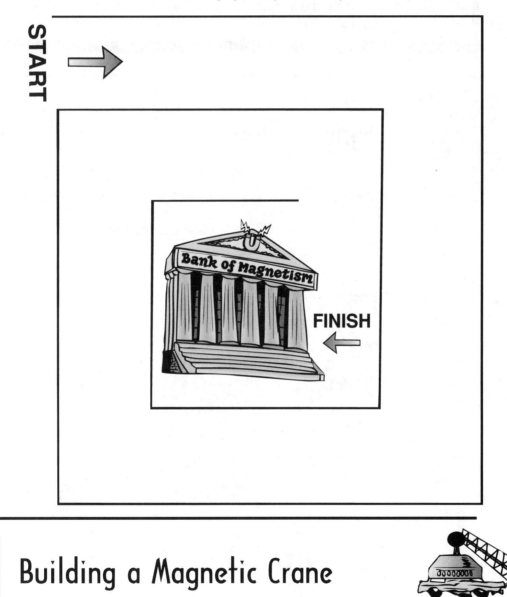

Building a Magnetic Crane

Somewhere in an auto wrecking yard, a large magnetic crane is moving cars and car parts around. The crane holds a large, powerful electromagnet. It is positioned on top of a car, and the magnet is electrified. The magnet attracts the car which is moved to where it is needed. The electricity is turned off and the car falls.

Name _____

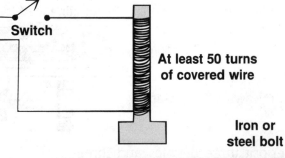

Switch

At least 50 turns of covered wire

Iron or steel bolt

Here is how to build your own magnetic crane. It won't pick up a car. It will pick up and move nails, bolts, washers, etc. The circuit is for the magnetic part. You will have to adapt a toy truck for movement.

Caution! Don't leave the switch closed too long. The wires may overheat and the battery could lose its strength.

 # Magnetic Play

Draw some small cartoon characters and attach paper clips to their base.

Write and act out a short play. Use a box as your theater stage as shown.

30

Name _____

Understanding Static Electricity

You and Static Electricity

Here are some examples of static electricity in your life.

1. A sweater crackles with electricity as you remove it.

2. Your hair stands on end after brushing.

3. Papers sometimes cling together.

4. You walk across a rug and get a shock as you touch a light switch.

5. You and a friend spark each other after walking across a rug.

6. Lightning due to static electricity frightens you.

7. Sliding across the car seat and touching the door handle.

Describe some of the static electricity experiences above that you have experienced. _____

_____

Have you had any experiences with static electricity besides those listed above?

_____

People have been interested in static electricity for a long time. An English scientist named Dr. William Watson developed this strange static experiment in the 18th century. A boy was suspended by silk ropes. A machine was cranked that rubbed the boy's feet. This created static electricity which the boy passed on to the girl's hand. The static electricity passed through her body and enabled her to attract bits of paper to her other hand.

Newton **does not** recommend that you try to duplicate this experiment.

Name _____

Newton Explains Static Electricity

There are two kinds of electricity. One is called **current** electricity. That is the kind that lights light bulbs and moves motors. The current is basically a stream of electrons running through copper wires.

Static electricity builds up on materials like glass or plastic that do **not** conduct electricity. The electrons involved in static electricity do not move freely. They are static which means stationary.

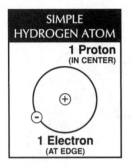

SIMPLE HYDROGEN ATOM
1 Proton (IN CENTER)
⊕
⊖
1 Electron (AT EDGE)

Every material in the universe is made of atoms. Atoms, in turn, are made of charged particles. The electrons on the outside of atoms have a **negative** electric charge. They have a minus sign symbol. ⊖ Protons inside the center of atoms have a **positive** electrical charge. They have a plus sign symbol. ⊕

Normal atoms have **equal** amounts of electrons and protons. They have no overall charge and are electrically **neutral**.

Static electricity always involves materials like wool, nylon, glass, rubber or plastic. When these non-conducting materials are rubbed, electrons can pile up on some and be removed from others. When electrons pile up on a rubbed object, it has **negative** ⊖ static electricity. If electrons are removed from a rubbed object, it has **positive** ⊕ static electricity. Protons rarely move.

Here's what happens to a plastic rod when it is rubbed with fur.

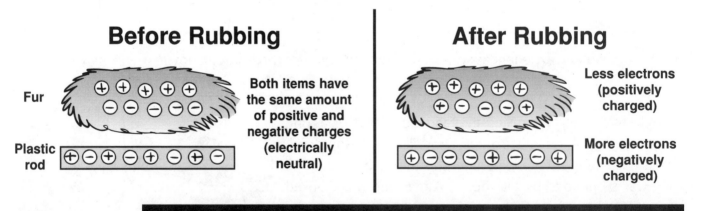

Before Rubbing

Fur

Plastic rod

Both items have the same amount of positive and negative charges (electrically neutral)

After Rubbing

Less electrons (positively charged)

More electrons (negatively charged)

Newton Note: Millions of electrons are involved in static electricity experiments. Only a few are shown in this diagram for simplicity.

This section on static electricity is difficult. Following are some review questions to help you understand the concepts.

Name _____

1. What are all materials in the universe made of? _____

2. Give the name and symbol of the two electrically charged parts of the atom.

   _____

   _____

3. Name some materials that **do not** conduct electricity and build up static electricity. _____

4. What is the difference between current and static electricity? _____

   _____

   _____

Static Electricity Rules and Laws

1. Static electricity experiments work best on dry days. Moisture in the air tends to leak electric charges.

2. Only electrons move in static experiments.

3. Static electricity is always due to rubbing or friction.

4. Normal objects are electrically neutral. To charge them, you must add or subtract electrons by rubbing.

5. Static electricity experiments only work with non-conductors such as glass and plastic.

6. To get a **positive** ⊕ static charge, you can rub glass with silk or nylon.

7. To get a **negative** ⊖ static charge, you can rub plastic with wool or fur.

8. Like charges repel each other.

9. Unlike charges attract.

Newton Note: Restudy the static electricity rules and laws described above. They will help you explain all the static experiments that you are about to do.

Plastic rods rubbed with wool

Glass rods rubbed with silk

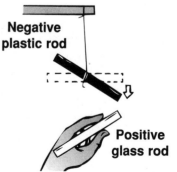

Negative plastic rod

Positive glass rod

Name _____

Experimenting with Static Electricity Rules

1. Obtain two large plastic combs, two glass rods or small test tubes, wool or fur, silk or nylon and some thin tape.

2. Use your tape to mount your plastic comb beneath a table as shown to the left. It should be balanced.

3. Rub your mounted comb **vigorously** with a wool strip.

4. **Quickly** rub another comb with wool and **slowly** bring it toward, but not touching, the mounted comb.

5. Review your static electricity rules.

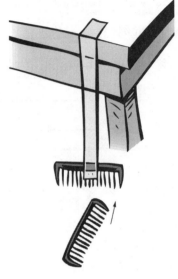

Describe what happened and why it happened. _____

_____

_____

6. Repeat the experiment using two glass rods or two small test tubes **vigorously** rubbed with silk or nylon. This is harder to do than the negative charge on the combs. Work fast and rub hard.

7. Review your static rules.

Describe what happened and why it happened. _____

_____

_____

8. Repeat using a comb mounted on a table. Charge the comb with wool. Quickly charge a glass rod or tube with silk or rayon. Let the two ends approach slowly.

9. Review your static rules.

Describe what happened and why it happened. _____

_____

_____

34

NEWTON'S
ACTION LAB
Electricity &
Magnetism
11

Static Electricity Experiments

Newton Reviews Static Electricity

Here is a review of the basic facts you will need to know to understand static electricity experiments.

1. The success of static electricity experiments varies considerably with the time of the year and especially with the air's humidity. On a dry day, static charges will be very good to you and obey all the rules. On a moist day, the charges would rather drift off into the atmosphere to play by themselves.

2. Static electricity is always due to rubbing or friction.

3. Only electrons move in static experiments.

4. Static experiments work primarily with non-conductors like glass or plastic.

5. To get a negative $\ominus$ charge, you rub wool or fur on plastic.

6. To get a positive $\oplus$ charge, you rub silk or nylon on glass.

7. Like charges repel $\ominus$ repels $\ominus$ $\oplus$ repels $\oplus$.

8. Unlike charges attract $\oplus$ attracts $\ominus$.

9. Charged objects attract neutral objects.

Static Pickup

1. Obtain a large comb or plastic rod and a piece of wool or fur. These will give you a strong negative charge.

2. Rub the comb and wool **vigorously**. Then try to attract various small objects around your room. Try anything available. Rerub the comb between each trial.

3. Name some of the small objects that you were able to attract. _____

_____

Name _____

4. Name some of the objects that you were **not** able to attract. _____

_____

Now for a paper pickup challenge.

5. Obtain a standard sheet of notebook paper. Cut it in half longways.

6. Charge your comb and try to pick the half sheet **completely** off the table.

7. If you don't succeed, cut 1" (2.5 cm) off of one end. Try to pick up the paper.

8. If you don't succeed, keep cutting inches (centimeters) off until you are able to completely lift the paper off the table. Give the size (length and width) of the

largest piece of paper you picked up. _____

Balloon Static Experiments

Balloons are non-conductors. They can be given a strong negative charge by rubbing them with fur or wool. Balloons keep these charges for a long time.

1. Blow up a balloon to the size of your head. Tie it so that the air can't escape.

2. Rub the balloon **vigorously** with fur or wool.

3. Slowly touch it to your or a friend's hair.

Describe what happened. _____

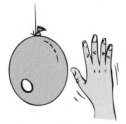

4. Rub vigorously and bring the balloon slowly toward hair on someone's head or arms.

Describe what happened. _____

5. Rub again and try to attach the balloon to a wall.

Describe what happened. _____

6. Attach about 20" (50 cm) of light thread to your balloon. Rub it and hold it suspended in the air.

7. Have a friend bring his or her hand **slowly** toward the charged balloon.

Describe what happened. _____

8. Recharge your balloon and have two friends compete to see whose hand is more attractive to the balloon. Their hands should slowly approach the balloon from opposite sides.

36

Name _____

9. Prediction time. Make your prediction **before** you try the experiment.

What would happen if you charged two balloons and **slowly** brought them

together? _____

_____

10. Now try the double balloon experiment.

What really happened? Can you explain what happened in terms of static laws?

_____

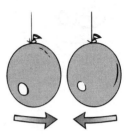

Static Balloon Freedom

This is your chance to be original. Use balloons and anything else available to develop your own balloon static experiments. Try to do at least two experiments using the outline below.

Balloon Static Freedom 1

a. What you did:

b. Sketch below:

c. Results:

Balloon Static Freedom 2

a. What you did:

b. Sketch below:

c. Results:

NEWTON'S ACTION LAB
Electricity & Magnetism
12

Static Cafeteria
(With a Charged Up Menu)

Newton Explains the Static Cafeteria

You cannot depend on static charges to work by the rules all the time. The air's humidity and the materials you can use vary widely. That's why this activity will be set up on a cafeteria basis.

You will be given a wide choice of activities. Select those that you readily have the materials for. You could use a plastic comb and fur or a balloon and wool to obtain a charge. Feel free to modify or improve any activity you wish.

Newton knows that you may run into problems. Console yourself with the thought that all scientists run into problems.

Newton Ruler Reminder: The plastic or balloon rubbed with wool or fur gives a negative ⊖ charge. Like charges repel. Unliked charges attract. A charged object will attract a neutral object.

Here are some items you may need for the static cafeteria. Decide which students can bring in which items.

string or thread	corks
aluminum foil	Ping-Pong™ balls
pins and nails	plastic wrap
balloons	Petri dishes
plastic comb	Styrofoam™
wool or fur	cereals (all kinds especially Cheerios™)
silk or nylon	paper or thin cards
glass or small test tubes	plus any odd items that you think static
plastic vials	charges will attract

Name _____

Static Experiments

Stubborn Paper

Tear a piece of notebook paper in half the longways. Put one half on top of the other on a table and rub vigorously with a pencil or pen.

What happened when you tried to separate the

papers? _____

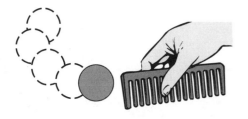

Ping-Pong™ Ball Attraction

Charge up a comb. Bring it slowly near a Ping-Pong™ ball resting on a table. It will follow you home.

What static rule does this demonstrate? _____

Charged Cereals

1. Place small cereal bits in a Petri dish or a small plastic vial. Rub the dish with silk or nylon.

Describe what happened. _____

_____

2. Try rubbing the dish and different cereal bits with wool.

Describe what happened. _____

_____

3. Tie a few inches (centimeters) of thread around an O-shaped cereal. Rub a comb vigorously with wool and bring it toward the cereal.

Describe what happened. _____

_____

4. Make a second suspended O-shaped cereal. Use a charged comb to touch both cereals. Now try to bring the two charged cereals together.

Describe what happened and explain why it happened. Hint: They both had the

same negative charge. _____

Name _____

Dancing Threads

Cut some sections of thin nylon and cotton thread. Bring a charged comb to the ends of the threads.

Describe what happened. Did the nylon and cotton threads act the same way?

_____

_____

The Static Compass

A regular compass responds to the Earth's magnetism. Let's build a static compass that responds to a static charge.

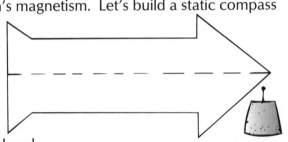

1. Cut out a paper arrow as shown.

2. Fold it along the dotted line.

3. Place a pin in a cork as shown.

4. Balance the folded arrow on the pin head.

5. Bring a charged comb toward (but not touching) the arrow. Move your comb **slowly** to the arrow.

Describe what happened. _____

_____

6. Newton Special. Try the same experiment, but this time let the charged comb **touch** the paper arrow. Then try to repel the arrow.

Describe your results. _____

_____

Static Cafeteria Freedom

In a real cafeteria, you get to pick the foods you wish to eat. In our static cafeteria, feel free to choose any of the materials around you. Combine them in any way to set up a static experiment that Newton would be proud of. Do at least three experiments using the form below.

Static Cafeteria Experiment Number _____

1. What do you plan to do?

2. Provide a labeled diagram of your experiment.

3. Describe your results.

40

Name _____

Super Static Experiments

Discovering Static Electricity

Static electricity was discovered and written about over 2500 years ago. A man named Thales was polishing a piece of amber jewelry. Amber is a plastic formed by hardened tree sap. Thales observed that the rubbed amber attracted dust and feathers.

It wasn't until 500 years ago that William Gilbert discovered that rubbed amber attracted many other objects. He called his discovery *electric* which came from the Greek word for *amber*.

Today there are still many things that you can discover about static electricity. Here are some interesting static experiments to get you started.

Attracting Water

1. Adjust a faucet to give a **slow thin** stream of water.

2. Rub a comb with wool or fur vigorously. A charged balloon works even better.

3. Bring the charge comb near **but not touching** the stream of water.

Describe what happened. Can you explain it in terms of the laws of static

electricity? _____

Charging a Fluorescent Light Bulb

1. Obtain a **small** fluorescent light bulb.

2. Go into a **darkened** room or closet.

3. Rub the light bulb vigorously with a piece of fur or wool.

Name _____

Describe what happened. _____

_____

Now let's try a variation on this experiment.

4. Obtain a blown up balloon.

5. Go back into the dark area.

6. Rub the balloon vigorously.

7. Touch the **end** of the fluorescent tube to the charged balloon.

Describe what happened. _____

_____

8. What atomic particle must have jumped from the balloon to the tube? _____

Newton Note: Be cautious with fluorescent tubes. A broken tube has danger-ous chemicals inside.

The Great Static Electricity Can Race

Metal cans conduct electrons. Yet they can be used in a simple static experiment.

1. Obtain a small, clean, empty fruit juice can. It should be about six fluid ounces (117 milliliters). Remove the label.

2. Place it on a table, charge a plastic comb and try to attract the can toward the comb. Don't touch the can.

Describe what happened. _____

_____

Newton Note: The comb has a negative charge. The area beneath the comb on the can becomes positive and they attract.

Name _____

Now for the static can race.

1. Obtain two fruit juice cans and two combs to charge. You might substitute a charged balloon for the comb.

2. Lay out a 2' (0.61 m) racetrack as shown.

3. Decorate your racing cans with felt pens to suit your personality. For a fair race, both cans must be the same size and the charged comb or balloon exactly alike.

4. Your can is eliminated if you **touch** it with your charged comb. May the best static electrician win!

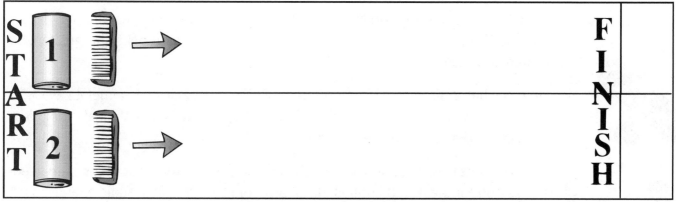

2 feet (0.61 m) long, start to finish

Name _____

Building an Electroscope

Newton Explains Electroscopes

Electroscopes are very useful instruments for detecting electrical charges. Electroscopes are based on the principle that like charges repel each other. If two thin metal foils are both given the same charge, they will repel. The greater the charge, the more they repel.

Here is what a laboratory electroscope looks like. It usually has a metal case with glass windows front and back. The thin metal usually used is gold leaf. Gold leaf reacts well to a charge but is very fragile.

Electroscopes have a round metal knob at the top where the static charge is placed. The reason for the round knob is that sharp points tend to leak static charges back to the air.

A charged object brought **near** the electroscope causes the two gold leaves to have the same charge. They repel each other and spread out. Remove the charged objects and the leaves return to normal. This process is called charging by **induction**.

If the knob is **touched** by the charged object, the charge remains on the leaves and they repel for a long time. They only go back to normal if the electric charge leaks into the air.

A charged electroscope can be discharged safely by placing your finger on the knob. The charges leave the metal foil as they pass into your body. It poses no danger to you as the static charge is weak.

How to Make a Simple Electroscope

There are many ways to build an electroscope. Build the best one you can with the materials available to you. A suggested model is shown on the following page.

Name _____

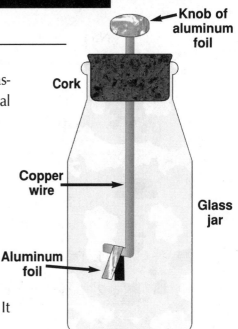

Knob of aluminum foil

Cork

Copper wire

Glass jar

Aluminum foil

1. Use as thick a copper wire as you can find. If your wire is plastic covered, just bare both ends. You can also use thin metal rods.

2. A metal ball is best at the top of your electroscope. A small metal doorknob works. You can also ball up aluminum foil to form your knob.

3. Most of you should use the **thinnest** aluminum foil you can find. You may find other thin foil. The foil can be bent in half as shown in the model.

4. The size of **each** half foil should be about 1 x 4 centimeters. It could be slightly bigger or smaller.

5. The jar could be metal or plastic. You could use a plastic see-through drinking cup. You could use any plastic container such as a clear, large medicine vial.

6. Go on to the next section when your electroscope is complete.

Foil Dimensions

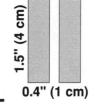

1.5" (4 cm)

0.4" (1 cm)

Testing Your Electroscope

1. Charge up a comb with wool or fur, and place it **near but not touching** the top of your electroscope.

Describe what happened. Use the word *induction* in your answer. _____

_____

2. Now touch the knob with your charged comb.

Describe what happened. _____

_____

3. Now touch your charged electroscope with your finger.

Describe what happened. Tell where the charge must have gone. _____

_____

Name _____

Electroscope Contest

People often talk about building a better mousetrap. Here is your chance to build a better electroscope. You will compete against teams of your classmates. The winning team will have an electroscope whose leaves **stay apart** the longest.

1. Form teams to share your ideas and material.

2. Build any variation of the electroscope that you think will win.

3. All electroscopes will compete on _____.

4. Bonus points for good workmanship.

You can charge your electroscope any way you wish. However, you are allowed only one contact with the electroscope top.

You are allowed three trials. Your best time before the leaves collapse will be recorded.

Your best time: _____ minutes _____ seconds

The winning time: _____ minutes _____ seconds

Name _____

Edison and Electricity

Newton Never Had Electricity

Sir Isaac Newton was born too soon. He never had electricity. The use of electricity in our homes, schools and factories only goes back 100 plus years. Poor Newton studied with candlelight and never heard a radio.

You take electricity for granted. It lights your home and pumps your fresh water. Here are some questions to make you appreciate electricity.

1. List at least 20 electrical devices in your home.

   1. _____   2. _____   3. _____   4. _____

   5. _____   6. _____   7. _____   8. _____

   9. _____   10. _____   11. _____   12. _____

   13. _____   14. _____   15. _____   16. _____

   17. _____   18. _____   19. _____   20. _____

2. Count the number of light bulbs in your home. Don't forget the refrigerator
   light. Don't forget the outside of your home. _____

3. Name the five **most important** electrical devices in your home.

   1. _____   2. _____   3. _____

   4. _____   5. _____

4. Name five electrical devices in your home that you could do **without**.

   1. _____   2. _____   3. _____

   4. _____   5. _____

Name _____

Thomas Alva Edison:
The World's Greatest Inventor

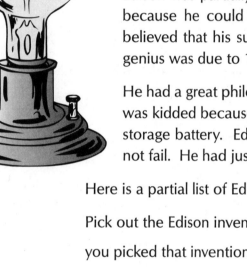

Edison was born in 1847 and died in 1931. He attended a real school for only three months in his entire life. Yet, in his lifetime, he was credited with over 1100 inventions.

Edison grew up extremely poor. At 12 years of age he sold candy and newspapers on trains. He even printed his own newspaper to sell on the train.

Edison was partially deaf most of his life. He considered deafness a benefit because he could read and work without hearing noisy distractions. He believed that his success was mostly due to hard work. It was said that his genius was due to 1% inspiration and 99% perspiration.

He had a great philosophy about not giving up on a science project. Once he was kidded because he failed thousands of times trying to perfect an electric storage battery. Edison looked at his failures differently. He said that he did not fail. He had just discovered thousands of ways that didn't work.

Here is a partial list of Edison's inventions. Most involved electricity.

Pick out the Edison invention that you think was his most important. Explain why you picked that invention. _____

_____

1. Incandescent light

2. Fluorescent light

3. Electric generators

4. Motion pictures

5. Phonograph

6. Duplicating machine

7. Improved the telephone

8. Discovered the basis for radios

9. Improved the typewriter

10. Automobile electronics

11. Improved the camera

12. Electric vacuum cleaner

13. Walking, talking dolls

14. High-speed printers

48

Name _____

Edison's Incandescent Light Bulb

Thomas Edison was one of many inventors trying to develop an **incandescent light** source. *Incandescent* means "something so hot that it gives off light." The sun for example, is incandescent. The moon is cold and merely reflects sunlight.

Edison, after years of research, found a filament that would give off light without burning up. He used carbonized filaments made from common cotton thread. His first filament lasted almost two days.

Edison improved his filament longevity by placing it in clear glass and pumping out the air. Later improvements involved adding argon gas to reduce the darkening of the inside glass.

Here is an experiment to illustrate incandescence.

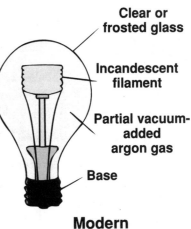

Clear or frosted glass

Incandescent filament

Partial vacuum-added argon gas

Base

Modern Incandescent Light Bulb

1. Obtain pliers and a small nail or pin.

Caution! The following should be done by an adult.

2. Use the pliers to hold the nail end over a Bunsen burner or gas stove flame for a few minutes as shown. **Be careful!** Place the hot nail in a safe place (ie. a glass of cool water) until it cools.

Describe the appearance of the hot nail end. It might help to darken the room.

_____

_____

Constructing a Model Light Bulb

Here is an experiment that tries to duplicate Edison's original incandescent light. Just like Edison, you may have to try many times to make it work.

1. Obtain a jar, a two-hole rubber stopper, a six-volt large battery, some covered copper wire, a simple switch and a **very thin** strand of copper or steel picture wire to act as a filament.

Name _____

Newton Caution! Adult supervision is recommended. Your wires will get hot.

2. Hook your materials up as shown but **leaving the switch open**.

3. The covered copper wires should be bare when connecting to the battery switch and inside the jar.

Newton Hint: Try finding wires with alligator clips attached. They make for easy electrical connections.

4. The filament wire must be **thin**. Wrap it tightly around the exposed wires inside the jar.

5. Darken the room and dim the lights.

6. Close the switch.

Describe your results, if any. _____

_____

7. If your filament doesn't light up, check connections. Try a better battery or a thinner filament wire.

Edison Makes History. Back in 1882 Edison developed the first electric power station in New York City. He had 500 customers using 10,000 of his light bulbs.

50

Name _____

Generating Electricity

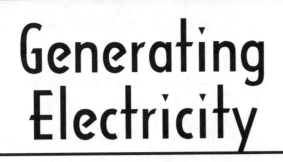

Newton Explains Electricity

Electricity is a wonderful energy source. It can be generated in many ways. It can be made in Nevada and delivered cheaply to Los Angeles. In your home, you can convert electricity to light, heat, sound, television pictures or the moving motor of a vacuum cleaner.

The graphs to the right explain how electricity is generated and used in the United States. Study the graphs and answer the questions.

How Our Electricity Is Generated

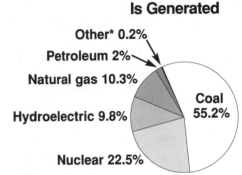

Other* 0.2%
Petroleum 2%
Natural gas 10.3%
Hydroelectric 9.8%
Nuclear 22.5%
Coal 55.2%

*Includes geothermal, wind, solar and others.

1. What energy source generates the most electricity? _____

2. What percent of our electricity is generated by nuclear energy? _____

3. *Hydro* means "water" in Latin. Can you explain the source of energy that is converted to hydroelectric power. _____

_____

4. Do homes, factories or stores consume the most electricity? _____

Where Our Electricity Is Used

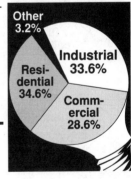

Other 3.2%
Industrial 33.6%
Residential 34.6%
Commercial 28.6%

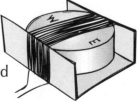

Electricity, Magnetism and Galvanometers

Electricity and magnetism are closely related. Magnets and wires in motion can generate electricity. Any wire carrying electricity has a magnetic field around it.

You can learn about this relationship by building a primitive **galvanometer**. You can use it to measure electricity.

1. Obtain stiff cardboard, a compass, several feet of **thin insulated** copper wire and a small $1^1/_2$-volt battery.

2. Bend and cut the cardboard so that it forms a two-sided case for the compass.

3. Wind 40 to 50 turns of wire around the box as shown. You should still be able to see the face of the compass from the side.

Name _____

To Galvanometer

4. Leave about 18" (46 cm) of wire sticking out from **both** ends of your coil. Use tape to hold the coil in place.

5. Remove the insulation from an inch (2.5 cm) of each end for connecting purposes.

6. Align your compass as shown on page 51 with east and west at the open sides. Tape the compass in this position. Rotate the entire galvanometer so that the compass needle is facing north.

7. Attach the two bare wires to the ends of your battery with tape.

Describe what happened. _____

8. Reverse the battery leads.

Describe what happened. _____

Save your galvanometer for the next section.

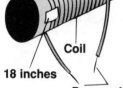

Turned by a turbine

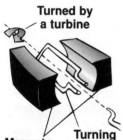

Magnets Turning coil of wire

Generating Electricity with Wire Coils, Magnets and Motion

Most electricity is made by moving a coil of wire inside a magnetic field. The motion of the coil can be generated by steam or falling water turning a turbine. A real generator has powerful magnets and massive coils.

Here is how you can make your own electricity.

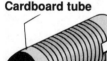

Cardboard tube

Coil

18 inches

Bare ends

1. Obtain a cardboard toilet paper tube, thin insulated copper wire, a bar magnet and an electric meter or your galvanometer.

2. Let 18" (46 cm) stick out and wind a tight coil around the cardboard tube. Leave another 18" (46 cm) sticking out at the other end. Use tape to hold the coil in place.

3. Bare the copper wire at each end to make connections.

4. Attach both bare ends to your electric meter or your compass galvanometer.

5. Move the magnet rapidly back and forth within your coil of wire.

Describe what happened. _____

_____

Electric meter or your galvanometer

Your simple coil and magnet generates very little electricity. What could you to

do generate more electricity? _____

Name _____

6. Place your magnet inside the coil. Do not move it. Was any electricity generated?

_____

7. Move the magnet in **quickly** but only **inward**. Then move it **quickly outward**. Observe the electric meter as you do this.

Describe what happened. _____

An Electrical Generating System

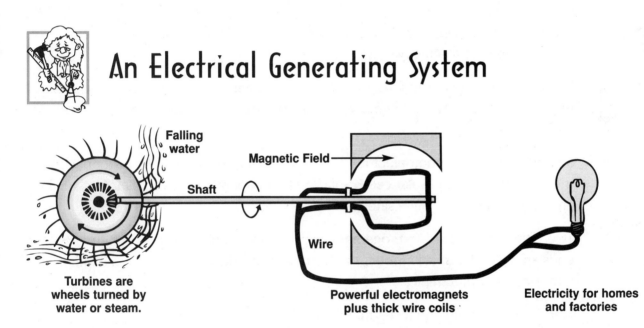

Falling water

Magnetic Field

Shaft

Wire

Turbines are wheels turned by water or steam.

Powerful electromagnets plus thick wire coils

Electricity for homes and factories

Most electricity is made like this. The power to turn the turbine can come from hydroelectric (falling water) or from steam heated by coal, oil, gas or nuclear energy.

There are other ways to generate electricity. They will be explained in the next unit.

NEWTON'S ACTION LAB
Electricity & Magnetism
17

More Ways to Generate Electricity

Electricity Without Magnets

Most electricity is generated using wires, magnets and motion. Here are some other ways to generate electricity.

Some materials such as quartz form crystals that can be twisted, bent or squeezed to produce electricity. Many microphones use these kinds of crystals to reproduce sound.

Some materials are **photoelectric**. When light shines on them, they produce electricity. They are called **solar cells**. Solar cells can operate your calculator or provide an energy source for space satellites.

Two **different** wires twisted together are called a **thermocouple**. When the twisted ends are heated, a small amount of electricity is produced. Your home heating system uses a thermocouple.

Here's how to build and test a thermocouple.

1. Obtain a thermocouple, if possible, from a heating and air conditioning company. If you can't obtain one, proceed to the next step.

2. Obtain a thick copper wire and a steel wire. You can use a coat hanger for the steel wire if you scrape the ends clean of paint.

3. Use pliers to **tightly** twist one bare end of each wire together.

4. Hook the other bare end to an electric meter or your homemade galvanometer. Remember to line up the needle with north.

5. **Carefully** hold with pliers and heat the twisted ends with a Bunsen burner or candle.

Describe your results. _____

_____

6. If time permits, build and test other thermocouples.

Name _____

Chemical Electricity

Cross-Section of a Typical Battery

Chemical reactions can generate electricity. Your automobile battery is a good example.

Chemicals making electricity are called a **cell**. Most common cells generate only $1\frac{1}{2}$ volts of electricity. A group of cells make up a battery. A six-volt battery is made up of four $1\frac{1}{2}$-volt cells.

All electric cells use two different materials separated by a fluid or moist area called an electrolyte. Two common materials are carbon and zinc.

Study the battery shown to the right. It is typical of most. Have an adult hacksaw a battery in half for you to observe.

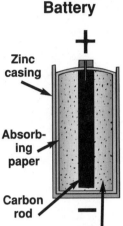

Zinc casing

Absorbing paper

Carbon rod

Electrolyte mixed with absorbent material

> **Caution!** Do not open an alkaline battery such as a Duracell™ battery.

Try to identify each part of the cut battery.

Some Homemade Batteries

An electric cell consists basically of two different metals separated by an electrolyte that can conduct electricity. Let's try to build an electric cell.

1. Obtain a lemon, a strip of copper and a large galvanized (zinc coated) nail.

2. Plunge the copper strip and the nail into the lemon about **1" (2.5 cm) apart**.

3. Attach wires to your copper and nail, and attach to an electric measuring device. You can use an electric meter or your galvanometer.

> **Newton Special:** You can also detect electricity using a portable radio earphone. It will click when your battery is attached to it.

Describe your results.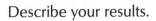

Name _____

Freedom Time

Lemons, copper and iron do make an electric cell. So do many other materials. Try placing a dime and a penny into a lemon. Try other materials such as aluminum. Try an orange, grapefruit, potato or even a dill pickle as an electrolyte. Describe and draw your cell combination below. Tell how it compares to the standard copper, nail and lemon electric cell.

_____

_____

_____

Electricity from Nuclear Power

Some states generate more than half their electricity using nuclear energy. In a nuclear power plant, uranium atoms split in a process called a **chain reaction**. The chain reaction releases large amounts of heat. The heat is used to boil water which turns to high energy steam. The steam turns a turbine which turns an electric generator.

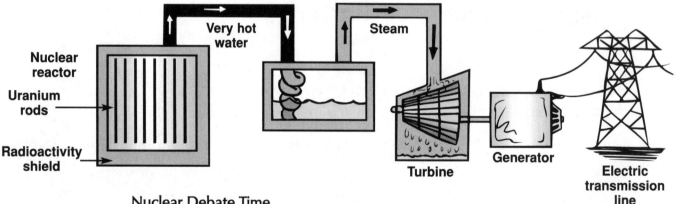

Nuclear Debate Time

Many people favor nuclear-powered electricity because it is clean, cheap, safe and available to us long after we run out of coal, oil and natural gas. Many people oppose nuclear power plants because they produce harmful radiation and may contaminate water supplies. A nuclear power plant is also more expensive to build than a regular electric plant.

What do you think? Form teams to debate this important energy issue. It would help to do some research on the topic. The nuclear debate is scheduled for _____. Make it a lively one.

Name _____

Electrical Circuits

Newton Never Had Electricity

Current electricity is a flow of electrons through a conducting circuit.
It is similar to a flow of water through a pipe. In an electric circuit, it
is electrons flowing through a wire.

Study the bell circuit to the right. It shows electrons flowing **out** of the
negative side of the battery. The electrons flow through the bell causing a ring-
ing. The electrons then flow **back** to the positive side of the battery.

The flowing out of a battery into a device like a bell or light and back to the battery is called a **complete circuit**. Nothing happens in an electric circuit if it is not
complete.

Electricians often use symbols instead of words for the different parts of a circuit.
Here are some common symbols.

Wire	Battery	Light	Switch (open)
	positive negative side (+) side (-)		

Sometimes you can't tell if you are connecting to the positive or negative end of a
battery. Here is a fun way to identify the positive end of a battery.

1. Obtain a potato, a battery and two copper wires.

2. Cut the potato in half.

3. Place the **bare** edge of two copper wires about 1" (2.5 cm) into the potato.

4. Connect the other two bare ends to any battery. One to each end. Tape
   the ends to the battery if necessary.

5. Wait three to five minutes.

6. Remove both wires.

Observe the two potato holes carefully. One is greenish. That was due to
the copper wire connected to the **positive** end of your battery.

Name _____

Electrical Conductors and Insulators

Some materials, such as copper, carry electrons easily. They are called **conductors**. Some materials, like glass, do not carry electrons easily. They are called **insulators**.

Let's test many common objects to find if they are conductors or insulators.

Alligator clips

1. Obtain a 1$\frac{1}{2}$-volt battery, a 1$\frac{1}{2}$-volt miniature light bulb with a matching base and some copper wires.

Newton Note: Electric circuits are easier to assemble if you obtain a Jump Cord Test Set. They have alligator clips at their ends for quick connection.

2. Hook up your materials to make the circuit shown.

3. Briefly touch the two test points together. The light should turn on.

Test points

Here is how your circuit works. If you place a conductor across the test points, the light will go on. If you place an insulator across the test points, no electricity will flow and the light bulb does not light up.

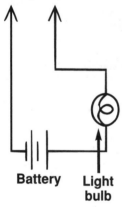

4. Make a collection of odds and ends. Use paper clips, nail files, combs, cork, rubber, coins, paper, chalk, your skin, pencils, hairpins, aluminum foil plus any objects within reach.

5. Fill out the Conductivity Data Table.

Battery   Light bulb

CONDUCTIVITY DATA TABLE			
Material Tested	**Results**		
	Good	Fair	Poor
Example: Plastic			✓
1.			
2.			
3.			
4.			
Put additional results on your own paper and attach.			

6. On the basis of your data table, make a general statement as to what types of material do or do not conduct electricity.

_____

_____

58

Name _____

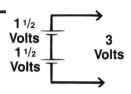

Simple Electric Circuits

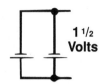

Batteries and light bulbs can be hooked up in a **series** or **parallel**.

Batteries hooked up in series adds the voltage together to give a higher voltage.

Batteries hooked up in parallel still have only 1½ volts total. They combine to make lights last longer.

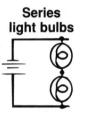

Two 1½-volt light bulbs hooked in series share the 1½ volts and may not light up without higher voltage. Some Christmas tree light bulbs are hooked up in series. If one goes out, they all go out.

Two 1½-volt light bulbs hooked in parallel can both light on a 1½-volt battery. If one burns out, the other remains lit. Your house electricity has parallel circuits.

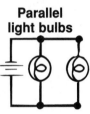

1. Obtain some batteries, light bulbs and bases and wire.

2. Try to hook up simple series and parallel circuits.

> **Newton Warning!** Be careful of shorting your batteries. Your wires can get very hot. Don't hook up 1½-volt lights to more than 1½-volt batteries. You can burn out the light.

Newton Explains a Flashlight Circuit

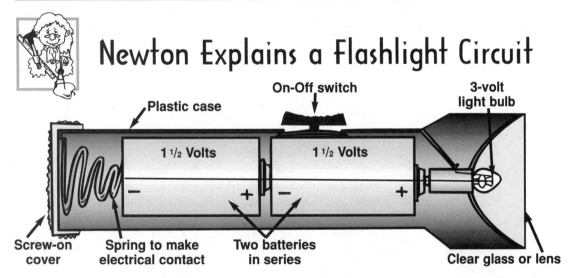

The two 1½-volt batteries are connected in series to give a total of three volts. The spring at the base makes contact with the battery and one end of the on-off switch. The light bulb connects to the positive end of the top battery as well as the other end of the switch. Turn the switch on to complete the circuit and your light goes on.

Name _____

Newton Puzzler Circuits

1. Which one of these lights will turn on when the switch is closed? _____

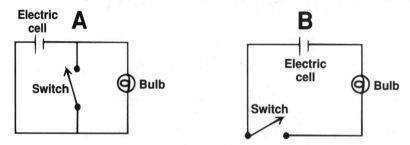

2. A bird is sitting on a high-voltage wire. Will the bird get an electric shock? _____

NEWTON'S
ACTION LAB
Electricity &
Magnetism
19

Fun with Electric Circuits

Electric Circuits in Your Life

Electric circuits run your radio, television and telephone. Electric circuits control cars, airplanes and satellites. Electric circuits make, control and distribute energy to homes and factories.

The electric circuits you are about to build are very useful but simple. Some electric circuits are built inside an **integrated circuit** chip. They can be the size of your thumbnail and be equal to millions of wires and switches.

Enjoy the simple circuits described in this unit. Think safety. Don't short the batteries. They can be ruined or get overheated.

There is one additional part that you must get acquainted with. It is called a double throw switch. Its picture and symbol are on the right.

Double throw switch

Symbol for double throw switch

Two-Switch Light Control

You may have this circuit in your home. A long hall may have a light switch at both ends. A stairway may have a light switch at the top and bottom. Here is what the circuit looks like. Can you build it using a miniature light bulb?

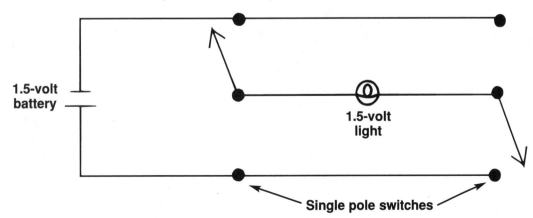

1.5-volt battery

1.5-volt light

Single pole switches

Name _____

Splitting Water

Water Molecule

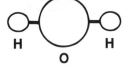

Water is a molecule that is made up of two hydrogen atoms and one oxygen atom electrically attached. It is given the symbol H_2O.

Electricity can be used to break the water molecule into oxygen gas and hydrogen gas. You will see the gas bubbles.

Hydrogen gas always appears at the negative connection. Oxygen gas appears only slightly at the positive connection.

Water-Splitting Circuit

+ −

6 volt

1. Obtain a glass or plastic container. Do not use a metal container. Obtain a six-volt battery, some covered copper wire and two large nails.

2. Fill the container about two-thirds full with water. Add about five tablespoons (25 ml) of vinegar. Vinegar is an acid that carries electricity. Add more vinegar if it works too slowly.

3. Strip both ends of your wire bare.

4. Wrap the two bare wires tightly around the large nails. Place the nails at the opposite sides of the jar as shown.

5. Connect the other **bare** ends to the six-volt battery.

6. Observe what forms at the wires.

What have you done to the water molecule? _____

_____

What gas appears on the negative wire? _____

Clean nails

Glass with water and vinegar

Copper-Plating a Nail

Electricity can be used to plate many objects. Your fine silverware was done with electric plating, so was the shiny bumper of your car. Following is a simple way to cover a nail with copper.

Name _____

1. Obtain a jar, such as a baby food jar, a six-inch volt battery, a strip of insulated copper wire about 6" x $^1/_2$" (15 x 1.5 cm), a large iron nail and some copper sulfate solution.

2. Connect your copper-plating circuit as shown below.

3. Connect the bared ends of a copper wire firmly to the nail and the copper strip.

4. Connect to the six-volt battery as shown. The **negative** must go to the nail.

5. Observe your nail after a few minutes.

You have now copper-plated a nail. You could improve your copper-plating using warm water or a longer plating period.

Now try to copper-plate other metal objects, such as keys, using the same copper sulfate solution. **Reminder.** Connect the object being plated to the negative side of the battery.

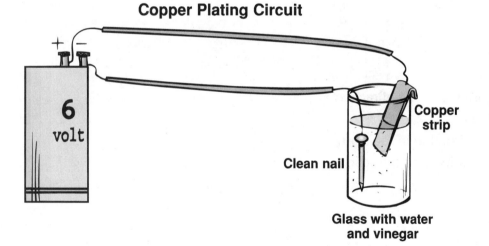

Copper Plating Circuit

Answer Key

Magnets Are Useful, page 6

1. radio	2. motors
3. compass	4. radar
5. bells	6. speakers
7. toys	8. meters
9. tape	10. computer

The Bar Magnet Puzzler, page 26

To solve this puzzler, you must know the nature of a magnet. A magnet has a north and south pole. The magnetic force is greatest at the poles and a minimum at the magnet's center. If you place either the north or south pole of the real magnet anywhere on the iron bar, you will feel attraction.

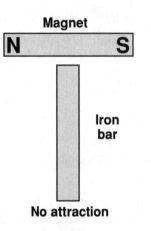

Magnet

N   **S**

Iron bar

No attraction

If you place the end of the iron bar in the center of the magnet, there will be little or no attraction.

Newton's Magnetic Car, page 26

The magnetic car will go nowhere. The forces pulling the car are equal and opposite. They cancel out and the car remains motionless.

Newton's Puzzler Circuits, page 60

Letter "B" shows the only complete circuit. The switch in "A" will short out the light bulb. The bird will not get shocked unless it **also** touches a second wire and completes the circuit.

64